U.S. ENVIRONMENTAL PROTECTION AGENCY
OFFICE OF INSPECTOR GENERAL

ANNUAL PLAN

Fiscal Year 2014

Scan this mobile
code to learn more
about the EPA OIG.

The Office of Inspector General (OIG) Annual Plan is produced by the OIG with input from the U.S. Environmental Protection Agency (EPA) Administrator, Deputy Administrator, Assistant Administrators and Regional Administrators, as well as congressional stakeholders and the Office of Management and Budget.

This plan is available in hard copy from:

Office of Inspector General
U.S. Environmental Protection Agency
MC 2491T
1200 Pennsylvania Avenue, NW
Washington, DC 20460

by calling (202) 566-2391

or via the Internet at: www.epa.gov/oig

Abbreviations

CFR	Code of Federal Regulation
CSB	U.S. Chemical Safety and Hazard Investigation Board
EPA	U.S. Environmental Protection Agency
FMFIA	Federal Managers' Financial Integrity Act
FTE	Full-Time Equivalent
FY	Fiscal Year
OI	Office of Investigations
OIG	Office of Inspector General
OMB	Office of Management and Budget

Hotline

To report fraud, waste or abuse, contact us through one of the following methods:

email:	OIG_Hotline@epa.gov
phone:	1-888-546-8740
fax:	1-202-566-2599
online:	http://www.epa.gov/oig/hotline.htm

write: EPA Inspector General Hotline
1200 Pennsylvania Avenue, NW
Mailcode 2431T
Washington, DC 20460

Suggestions for Audits or Evaluations

To make suggestions for audits or evaluations, contact us through one of the following methods:

email:	OIG_WEBCOMMENTS@epa.gov
phone:	1-202-566-2391
fax:	1-202-566-2599
online:	http://www.epa.gov/oig/contact.html#Full_Info

write: EPA Inspector General
1200 Pennsylvania Avenue, NW
Mailcode 2410T
Washington, DC 20460

Message from the Inspector General

I am pleased to present the U.S. Environmental Protection Agency (EPA) Office of Inspector General (OIG) Annual Plan for fiscal year (FY) 2014. This document describes how the OIG will promote economy, efficiency and effectiveness, and prevent and detect fraud, waste and abuse, through independent oversight of the programs and operations of the EPA and the U.S. Chemical Safety and Hazard Investigation Board. This plan reflects the priority work that the OIG believes is necessary to keep the Administrator and Congress fully informed about problems and deficiencies relating to the administration of agency programs and operations.

Arthur A. Elkins Jr.

This OIG Annual Plan identifies mandated and selected assignment topics continuing from FY 2013 and scheduled to start during FY 2014. Although this plan provides a framework for activities we intend to carry out in FY 2014, the OIG often undertakes unanticipated work based on legislative mandates, congressional inquiries, hotline requests or governmentwide reviews.

Our plan is implemented through audits, evaluations, investigations and follow-up reviews in compliance with the Inspector General Act, the applicable professional standards of the Comptroller General of the United States, and the Quality Standards for Federal Offices of Inspector General of the Council of the Inspectors General on Integrity and Efficiency. Readers are encouraged to consult our website, www.epa.gov/oig, for the most current listing of recently issued reports relating to our implementation of the plan.

Primary sources of input for the assignments listed in this plan included risk assessments across agency programs and operations based upon prior OIG work, U.S. Government Accountability Office high-risk assessments, congressional interest, Office of Management and Budget (OMB) priorities, agency vulnerability/internal control assessments under OMB Circular A-123 and the Federal Managers' Financial Integrity Act, and identification of key agency challenges and strategic planning priorities. Our current planning also reflects direct outreach and solicitation of topics and assignment suggestions from EPA's leadership and external stakeholders (see appendix B). Other assignments are required or are self-initiated based upon our strategic themes, which are focused on providing the greatest value and risk reduction to the agency and the greatest benefit to public health.

We want to thank each member of the agency leadership, as well as external stakeholders and our staff, for their direct participation in this process. We look forward to continuing an open dialogue for receiving their ideas, suggestions and feedback. We welcome input into our planning process and feedback on the quality and value of OIG products and services from all customers, clients, stakeholders and the public via webcomments.oig@epa.gov.

Arthur A. Elkins Jr.
Inspector General

Table of Contents

About the EPA Office of Inspector General

EPA Office of Inspector General

The Office of Inspector General (OIG) is an independent office of the U.S. Environmental Protection Agency (EPA) that promotes economy, efficiency and effectiveness, and prevents and detects fraud, waste and abuse, through independent oversight of the programs and operations of the EPA and the U.S. Chemical Safety and Hazard Investigation Board (CSB).

The EPA OIG was created and is governed by the Inspector General Act of 1978 as amended (5 U.S.C. App. 3). The Act established offices of Inspector General as independent and objective units to:

1. Conduct and supervise audits and investigations relating to the programs and operations of their agencies.
2. Review existing and proposed legislation and regulations relating to the programs and operations of their agencies.
3. Provide leadership and coordination, and recommend policies for activities designed to promote economy, efficiency and effectiveness, and to prevent and detect fraud and abuse.
4. Provide a means for keeping the head of the establishment and Congress fully and currently informed about problems and deficiencies, and the necessity for any progress of corrective actions.

EPA OIG staff members are physically located at headquarters in Washington, D.C.; at regional headquarters offices for all 10 EPA regions; and at other EPA locations including Research Triangle Park, North Carolina, and Cincinnati, Ohio.

In fiscal year (FY) 2004, Congress designated the EPA Inspector General to also serve as the Inspector General for the CSB.

EPA's Mission

The EPA's mission is to protect human health and the environment. The OIG Strategic and Annual Plans are specifically designed to connect implementation of the Inspector General Act with the EPA's mission for the most economical, efficient and effective achievement of the EPA's performance goals. In appendix A, we provide more details about our FY 2014 annual performance measures and targets. The list below identifies the EPA's strategic goals, cross-cutting fundamental strategies and priority themes that we take into account when planning audits, evaluations and investigations.

EPA's FY 2011–2015 Strategic Goals, Cross-Cutting Strategies and Priority Themes
EPA's Strategic Goals
• **Taking Action on Climate Change and Improving Air Quality**   *Protect and improve the air so it is healthy to breathe, and risks to human health and the environment are reduced.*
• **Protecting America's Waters**   *Protect and restore our waters to ensure that drinking water is safe, and that aquatic ecosystems sustain fish, plants and wildlife, and economic, recreational, and subsistence activities.*
• **Cleaning Up Communities and Advancing Sustainable Development**   *Promote sustainable, healthier communities and protect vulnerable populations and tribal communities.*
• **Ensuring the Safety of Chemicals and Preventing Pollution**   *Ensure the safety of chemicals that are used in consumer products, the workplace, and the environment.*
• **Enforcing Environmental Laws**   *Protect human health and the environment through vigorous and targeted civil and criminal enforcement.*
EPA's Cross-Cutting Fundamental Strategies
• **Expanding the Conversation on Environmentalism**
• **Working for Environmental Justice and Children's Health**
• **Advancing Science, Research, and Technological Innovation**
• **Strengthening State, Tribal, and International Partnerships**
• **Strengthening the EPA's Workforce and Capabilities**
EPA's Priority Themes
• **Making a Visible Difference in Communities across the Country**
• **Addressing Climate Change and Improving Air Quality**
• **Taking Action on Toxics and Chemical Safety**
• **Protecting Water: A Precious, Limited Resource**
• **Launching a New Era of State, Tribal and Local Partnerships**
• **Embracing EPA as a High Performing Organization**
• **Working Toward a Sustainable Future**

Strategic Plan Outline

Vision

Be the best in public service and oversight for a better environment tomorrow.

Mission

Promote economy, efficiency, effectiveness, and prevent and detect fraud, waste, and abuse through independent oversight of the programs and operations of the Environmental Protection Agency and Chemical Safety and Hazard Investigation Board.

Goals

1 Contribute to improved human health, safety and the environment

2 Contribute to improved EPA and CSB business practices and accountability

3 Be responsible stewards of taxpayer dollars

4 Be the best in public service

Objectives

- Influence programmatic and systemic changes and actions that contribute to improved human health, safety and environmental quality
- Add to and apply knowledge that contributes to reducing or eliminating environmental and infrastructure security risks and challenges
- Make recommendations to improve EPA and CSB programs

- Influence actions that improve operational efficiency and accountability, and achieve monetary savings
- Improve operational integrity and reduce risk of loss by detecting and preventing fraud, waste, abuse or breach of security
- Identify best practices, risks, weaknesses and monetary benefits to make recommendations for operational improvements

- Promote and maintain an accountable, results-oriented culture
- Ensure our products and services are timely, responsive, relevant, and provide value to our customers and stakeholders
- Align and apply our resources to maximize return on investment
- Ensure our processes and actions are cost effective and transparent

- Maintain the highest ethical standards
- Promote and maintain a diverse workforce that is valued, appreciated and respected
- Enhance constructive relationships and foster collaborative solutions
- Provide leadership, training and technology to develop an innovative and accomplished workforce

Identifying the Risks

As required by the Reports Consolidation Act of 2000, the OIG reviewed the major risks, challenges and planning priorities across EPA and solicited first-hand input from agency leadership to identify and select OIG products and topics that would be of greatest benefit to the agency and the American public it serves. This section summarizes and applies the key FY 2013 agencywide risks, issues and management challenges that help guide the general direction and focus of OIG audits, evaluations and investigative work.

Top EPA Management Challenges—Reported by the OIG for FY 2013

1. **Oversight of Delegations to States**: Due to differences between state and federal policies, interpretations, strategies and priorities, the EPA needs to more consistently and effectively oversee its delegation of programs to the states, assuring that delegated programs are achieving their intended goals.

2. **Safe Reuse of Contaminated Sites**: The EPA's duty is to ensure that reused contaminated sites are safe for humans and the environment. The EPA must strengthen oversight of the long-term safety of sites, particularly within a regulatory structure in which non-EPA parties have key responsibilities, site risks change over time, and all sources of contamination may not be removed.

3. **Enhancing Information Technology Security to Combat Cyber Threats**: The EPA is highly vulnerable to existing external network threats, despite reports from security experts that Advanced Persistent Threats, designed to steal or modify information without detection, are becoming more prevalent throughout the government.

4. **EPA's Framework for Assessing and Managing Chemical Risks**: The EPA's effectiveness in assessing and managing chemical risks is limited by its authority to regulate chemicals under the Toxic Substances Control Act. Chemicals manufactured before 1976 were not required to develop and produce data on toxicity and exposure, which are needed to properly and fully assess potential risks.

5. **Workforce Planning**: The EPA's human capital is an internal control weakness in part due to requirements released under the President's Management Agenda. The EPA has not developed analytical methods, and does not collect data needed to measure its workload and the corresponding workforce levels necessary to carry out that workload.

EPA Internal Control Risks and Weaknesses Identified by the OIG for FY 2013

We identified the following EPA internal control weaknesses as part of our annual Federal Managers' Financial Integrity Act (FMFIA) activities.

- Tribal Environmental Capacity Building

- Recovery of Funds

- Contract Management

- Compass System Limitation Area Material Weakness to the EPA's Account Operations

Risks, Priorities and Issues Identified by EPA Through OIG Stakeholder Outreach Interviews

The following identifies cross-cutting risks, priorities and issues identified through outreach solicitations and meetings with EPA leadership. In appendix B, we provide further details.

- Emergency Preparedness/Homeland Security

- Better Collaboration/Coordination with States and Other Federal Agencies with Environmental Mission and Authority

- Limitations of EPA Authority

- Consistent and Reliable Data and Performance Measurement

- Improving EPA Organizational Design and Coordination of Resources to Eliminate Duplication

- Monitoring of States, Grants Management, Compliance and Enforcement (How Much Delegation? Federal vs. State Roles?)

- Human Capital Management—Skill Gaps/Alignment With Functions and Workforce Restructuring

- Better Use of Technology, Information and Research

- EPA's Regulatory Process (Better and Faster Analysis of Costs, Science and Benefits)

- Cross-Media Risk Assessment, Planning and Priority Setting for Better Application of Resources

- Hydraulic Fracturing, Water Infrastructure, Financing and Water Availability

- Climate Change and Air

- Brownfields/Environmental Justice, Tribal Capacity

Annual Plan Strategy

Annual planning is a dynamic process and requires adjustments throughout the year to meet priorities and to anticipate and respond to emerging issues with the resources available. The OIG examines the cross-agency risk assessment, agency challenges, prior work, future priorities and customer input to develop and prioritize its FY 2014 work.

Making Choices—A Customer-Driven Process

OIG work that is not otherwise mandated is proposed, considered and selected through a rigorous process using the criteria listed below to develop a portfolio of assignments that represents the best possible return on investment in terms of monetary or public value and responsiveness in addressing the needs, risks, challenges, priorities and opportunities of OIG customers, clients and stakeholders. We conducted considerable outreach to agency leaders and stakeholders on environmental and management risks, challenges and opportunities. We conducted a risk assessment based upon previously identified risks and challenges. We invited our entire staff to formulate assignment suggestions from their immediate knowledge of EPA operations and the consideration of stakeholder input and risks.

Criteria Considered in Identifying and Selecting Audit and Evaluation Assignments for FY 2014

Environmental/Human Health/Business Risks Addressed, Including:
- What is the known extent of the issue (i.e., sensitive or other populations impacted, area involved, and environmental justice)?
- What are the potential environmental or human health benefits (return on investment) to be derived and the reduction or prevention of environmental, human health or business risks?

Potential Risk of Fraud, Waste or Abuse:
- What resources and data, physical or cyber security equipment, and program integrity and violations of laws/regulations are involved?

Opportunity for Improved Business Systems/Accountability, Including:
- How does the project align with the EPA's strategic goals/objectives?
- What is the expected return on investment (for example, potential questioned costs, funds put to better use or other potential monetary benefits, improved decision-making, improved data quality/reliability, reduced vulnerabilities, and strengthened internal controls)?

EPA Dollar/Full-Time Equivalent (FTE) Investment/Financial Impact (in relation to the EPA's overall resource level):
- What headquarters and regional resources are committed to the program, including FTEs?
- What resources are used including contracts, grants, state programs or other mechanisms, such as state funding, to accomplish the goals? How might this impact the program's implementation? What percentage of the program's funding is coming from state, other federal or private partnership resources?

Prior Audit/Evaluation Results:
- What are the conditions or changes since prior review by EPA OIG, U.S. Government Accountability Office or other auditing body?
- What new information or indications of auditable issues are available?

Stakeholder/Public Interest:
- Is the topic of the project generating interest from Congress, the public and news organizations? What is the interest and why?
- Who are the expected users of the project's product? How would it be used?

The Plan: Continuing and New Assignments for FY 2014

Office of Audit

OIG audit work focuses on five areas, with emphasis on identifying opportunities for cost savings and reducing risk of resource loss. Funds awarded for assistance agreements and contracts account for approximately two-thirds of the EPA's budget. Producing timely and reliable financial statements remains a priority across the federal government. Equally important is the need to gather, protect and use financial and program performance information to improve the EPA's accountability and program operations. The Office of Audit's five product lines are:

- Contracts and Assistance Agreement Audits.
- Efficiency Audits.
- Forensic Audits.
- Financial Audits.
- Information Resources Management Audits.

Specific assignments are listed on the following pages and will emphasize:

- Direct testing for fraud in grants, contracts and operational activities.
- Cost savings resulting from audits of grantee and contractor claims.
- Continued improvements in assistance agreements and contract administration.
- EPA's preparation of timely, informative financial statements.
- EPA's use of financial and program performance information, including efficiency measures, to identify cost savings and potential cost recoveries, reduce risks and maximize results achieved from its environmental programs.
- Reviews of the EPA's internal controls, including its risk assessment processes and allocation/application of human resources.
- The EPA's integrity of data and system controls, as well as compliance with a variety of federal information security laws and requirements, to ensure system and data integrity.

Following are definitions of OIG carryover, discretionary and mandated assignments:

- **Carryover Assignments**: Assignments still in progress that started in a prior fiscal year.
- **Discretionary Assignments**: Assignments designed to identify and prioritize projects in areas of highest risk that support OIG mission.
- **Mandated Assignments**: Assignments that the OIG is required to conduct by law or regulation.

Contracts and Assistance Agreement Audits

The Contracts and Assistance Agreement Audits product line is responsible for conducting performance audits of EPA's management of contracts, grants, cooperative agreements and interagency agreements.

Point of Contact: Janet Kasper (312) 886-3059

Title	Primary Objective	Estimated/Actual Start Date
Carryover		
Emergency and Rapid Response Service Contracts	To determine whether the EPA is effectively managing task orders under Emergency and Rapid Response Service Contracts.	October 2012
Reviews of Agency Purchase Card and Convenience Check Program	To determine whether the EPA has sufficient controls to identify illegal, improper and erroneous use of purchase cards.	December 2012
State Revolving Fund – Pace of Expenditures	Have the EPA and state actions to address large balances of Drinking Water State Revolving Fund Unliquidated obligations reduced such obligations?	March 2013
Puget Sound Action Agenda	To determine whether the EPA ensures that its grantees are effectively administering Puget Sound grants throughout the life of the grants.	June 2013
Discretionary		
Grant Advanced Monitoring	To determine whether the EPA's advanced administrative monitoring system is effective at ensuring that grant recipient costs are allowable.	October 2013
EPA's Contract Management Assessment Program	To determine whether assessments are sufficient to identify weaknesses in internal controls or systemic vulnerabilities.	October 2013
Strategic Sourcing of Contracts	To evaluate whether the EPA is using and gaining efficiencies from federal strategic sourcing initiatives.	January 2014
Hurricane Sandy Funding	To determine whether the Office of Water and Region 2 are effectively overseeing the Disaster Relief Act funding.	April 2014
Mandated		
Improper Payments Act Compliance for FY 2013	To determine EPA compliance with the 2010 Improper Payments Elimination and Recovery Improvement Act.	November 2013

Efficiency Audits

The Efficiency Audits product line is responsible for identifying ways for EPA programs and operations to improve processes and realize cost savings, thus freeing resources for high priority environmental projects.

Point of Contact: Mike Davis (513) 487-2363

Title	Primary Objective	Estimated/Actual Start Date
Carryover		
Controls for Travel of EPA Employees	To determine the effectiveness of EPA oversight and controls for employees in travel status.	April 2013
Management and Disposal of Underutilized Personal Property Stored in Warehouse Spaces	To determine the extent to which the EPA's personal property stored in select warehouse spaces are effectively utilized, accounted for and disposed of by the EPA.	April 2013
EPA's Fleet Management	To determine whether the EPA's fleet program is in accordance with the federal fleet requirements for utilization and fuel energy conservation.	June 2013
Discretionary		
Oversight of Guam, American Samoa, Commonwealth of the Northern Mariana Islands and U.S. Virgin Islands	To examine whether the EPA has controls and processes in place to ensure proper oversight of Guam, American Samoa, Commonwealth of the Northern Mariana Islands and U.S. Virgin Islands.	January 2014
EPA Investments in Information Technology Products and Services	To determine whether information technology investments in the EPA's Office of Environmental Information are efficiently and effectively managed to meet the agency's strategic goals and mission.	January 2014
Reliability of EPA Personal and Real Property Information	To determine whether the EPA has accurate and reliable information on its personal and real property to effectively manage the property and reduce environmental impact.	September 2014

Forensic Audits

The Forensic Audits product line is responsible for conducting financial audits of EPA assistance agreements and contracts to identify potentially fraudulent actions and determine the acceptability of costs claimed under specific financial instruments.

Point of Contact: Robert Adachi (415) 947-4537

Title	Primary Objective	Estimated/Actual Start Date
Carryover		
Pegasus Technical Services Inc. Contract	To determine whether costs charged to the contract are allowable, allocable and reasonable in accordance with contract terms and applicable government regulations.	October 2012
Cooperative Agreement #83456201 Awarded to the National Association of State Department of Agriculture Research Foundation	To determine whether cooperative agreement awards were conducted in accordance with 40 Code of Federal Regulations (CFR) Part 30 and costs incurred were allowable.	January 2013
Apex Logistics Contract	To examine whether costs charged to the contract are allowable, allocable and reasonable in accordance with contract terms and applicable government regulations.	January 2013
Cooperative Agreement #00T13801 Awarded to the California Air Resources Board	To determine whether cooperative agreement awards were conducted in accordance with 40 CFR Part 31 and costs incurred were allowable.	March 2013
Dozier Technologies Inc. Contract (EP-C-08-020)	To determine whether costs charged to the contract are allowable, allocable and reasonable in accordance with contract terms and applicable government regulations.	March 2013
Discretionary		
Brownfields Revolving Loan Fund Assistance Agreements	To perform assistance agreement audits of selected Brownfields recipients.	October 2013
Construction Grants Awarded to the District of Columbia Water and Sewer Authority	To determine whether the costs claimed under the grants are reasonable, allocable and allowable.	June 2014
Environmental Science and Engineering Fellowship Program Assistance Agreements	To perform assistance agreement audits of selected recipients to determine whether funds were expended in accordance with federal regulations.	March 2014
Review of Post-American Recovery and Reinvestment Act Diesel Emission Recovery Act Grants	To review Diesel Emission Recovery Act grants awarded.	September 2014
Forensic Review of Hotline Complaints	To review hotline complaints submitted to the OIG Hotline.	January 2014

Title	Primary Objective	Estimated/Actual Start Date
Mandated		
FY 2014 Single Audit Program	To review and process Single Audit reports that are prepared by Certified Public Accountant firms under the Single Audit Act.	January 2014
Region 9 Request – Review of Selected Tribes in Nevada	To review EPA grants open or closed within the past 3 years involving select tribes in Nevada.	October 2014

Financial Audits

The Financial Audits product line is responsible for rendering opinions on financial statements produced by the EPA, and also conducts performance audits of EPA financial matters for efficiency and effectiveness.

Point of Contact: Paul Curtis (202) 566-2523

Title	Primary Objective	Estimated/Actual Start Date
Carryover		
FY 2012 Financial Statements: Pesticides Reregistration and Expedited Processing Fund	To render an opinion on the agency's statements, and determine compliance with laws and regulations.	November 2012
FY 2012 Financial Statements: Pesticides Registration Fund	To render an opinion on the agency's statements, and determine compliance with laws and regulations.	November 2012
EPA's Accounts Receivable Internal Controls	To determine whether the EPA's accounts receivable internal controls function is effective and ensures the reliability of financial reports.	January 2013
EPA Biennial Use Fee Reviews	To determine effectiveness of the EPA's biennial use fee reviews.	January 2013
FY 2013 EPA Financial Statements	To determine whether the EPA's consolidated financial statements were fairly stated in all material respects.	April 2013
Discretionary		
Review of Unliquidated Obligations at Research Triangle Park	To determine whether the EPA has adequate controls in place to identify and deobligate unneeded contract and miscellaneous obligations.	March 2014
Review of Independent Government Cost Estimates and Indirect Costs for EPA's Interagency Agreements	To determine whether the EPA promotes sound financial practices.	March 2014
Mandated		
FY 2013 Financial Statements: Pesticide Reregistration and Expedited Processing Fund	To render an opinion on the agency's statements, and determine compliance with laws and regulations.	March 2014
FY 2013 Financial Statements: Pesticide Registration Fund	To render an opinion on the agency's statements, and determine compliance with laws and regulations.	March 2014
FY 2014 EPA Financial Statements	To determine whether the EPA's consolidated financial statements were fairly stated in all material respects.	June 2014

Information Resources Management Audits

The Information Resources Management Audits product line reviews the economy, efficiency and effectiveness of the agency's investments in systems for achieving environmental goals and ensuring integrity of data used for decision making; and reviews strategies for setting priorities, developing plans to accomplish the priorities, and measuring performance.

Point of Contact: Rudolph Brevard (202) 566-0893

Title	Primary Objective	Estimated/Actual Start Date
Carryover		
Implementation of Cross-Media Electronic Reporting Rule	To determine to what extent the EPA implemented a management control structure for the implementation of the Cross-Media Electronic Reporting Regulation.	January 2012
Skills Assessment of Personnel With Critical Information Security Responsibilities (Contracted)	To determine the effectiveness of the qualifications and current skills of EPA personnel with significant information security responsibilities.	February 2012
Review of EPA's Cloud Computer Initiative	To determine whether the EPA has adequately planned to migrate to the cloud.	October 2012
EPA's Controls Over Sensitive Personally Identifiable Information	To determine whether the EPA has implemented procedures and processes for protecting Personally Identifiable Information in accordance with federal and agency criteria.	November 2012
FY 2013 Federal Information Management Security Act Audit	To determine whether the EPA's Computer Security Program is comprehensive and actively implemented throughout the agency to balance risk and mission requirements.	April 2013
Mandated		
FY 2014 Federal Information Security Management Act Audit	To determine whether the EPA's Computer Security Program is comprehensive and actively implemented throughout the agency to balance risk and mission.	April 2013
Follow-Up on Significant Information Technology Security Findings	To determine whether EPA has implemented corrective actions to address significant information technology security findings.	March 2014
Status of Cloud-Computing Environments within the Federal Government	To determine each selected agency's efforts to adopt cloud-computing technologies.	March 2014

Title	Primary Objective	Estimated/Actual Start Date
EPA FY 2014 FISMA Audit	To conduct an independent audit of the agency's compliance with the Federal Information Security Management Act.	March 2014
Data Quality Review of Self-Reported Information in EPA's XACTA	To determine whether the EPA implemented management control processes for maintaining the quality of data in the EPA's XACTA System.	March 2014

Office of Program Evaluation

The Office of Program Evaluation examines root causes, effects and opportunities leading to conclusions and recommendations that influence systemic changes and contribute to the accomplishment of the agency's mission. Program evaluations answer questions about how well a program or activity is designed, implemented or operating in achieving EPA goals. Program evaluations may produce conclusions about the value, merits or worth of programs or activities. The results of program evaluations can be used to improve the operations of EPA programs and activities, sustain best practices and effective operations, and facilitate accomplishment of EPA goals. Evaluations by the Office of Program Evaluation are performed by staff with diverse backgrounds, including accounting, economics, environmental management and the sciences, and they comply with *Government Auditing Standards*.

Evaluation topics and priorities in our plan are driven by our assessment of organizational risk in relation to available resources and based on input from the EPA's leadership, Congress and stakeholders. Program evaluations are conducted by the following six product lines:

- Air.
- Water.
- Land Cleanup and Waste Management.
- Toxics, Chemical Management and Pollution Prevention.
- Science, Research and Management and Integrity.
- Special Program Reviews.

Assignments concentrate on all of the OIG themes, reflecting our attention to the agency's mission as well as the agency's operational and systemic risks. Specific assignment titles are listed on the following pages.

Air

The Air product line is responsible for conducting evaluations to assess EPA management of risks to provide reasonable assurance of progress toward goals and adequate protection to the public.

Point of Contact: Rick Beusse (919) 541-5747

Title	Primary Objective	Estimated/Actual Start Date
Carryover		
Use of Remote Sensing Data to Assess Contamination at Delisted Superfund Sites – Phase 2	To determine whether hyperspectral imaging data can be used to assess pollution concentrations in vegetation as a potential indication of pollutant concentrations at delisted Superfund sites.	October 2011
EPA Oversight of Clean Air Act Title V Fees	To determine whether the EPA's oversight of state and local Title V programs' fee revenue practices are effective in identifying and obtaining corrective actions.	June 2012
Congressional Request: Evaluation of EPA Office of Research and Development's Research on Human Subjects	To determine whether the EPA followed applicable laws, regulations, policies, procedures and guidance when it exposed human subjects to diesel exhaust emissions or concentrated airborne particles.	October 2012
Assessment of EPA Efforts To Reduce Methane Product Emissions From Leaking Pipes	To determine the effectiveness of the EPA's greenhouse gases requirements in addressing methane emissions.	June 2013
Discretionary		
EPA Region 2 Oversight of U.S. Virgin Islands Authorized Environmental Program	To determine whether the U.S. Virgin Islands implemented its EPA-authorized environmental program.	October 2013
Review of Enforcement Decree Compliance for Selected Clean Air Act Sources	To determine whether the EPA ensured that selected facilities with Clean Air Act violations comply with terms of their enforcement agreement.	October 2013
Review of Selected Ambient Air Quality Monitoring Networks	To assess whether the EPA effectively used annual network reviews to determine how well the monitoring network is achieving its objectives.	November 2013
EPA Efforts to Incorporate Environmental Justice into Clean Air Act Inspections for Air Toxics	To determine whether the EPA has targeted overburdened communities or communities with disproportionate impacts for air toxics inspections.	December 2013

Water

The Water product line is responsible for conducting evaluations to assess the EPA's protection and restoration of healthy aquatic communities and waters that sustain human health.

Point of Contact: Dan Engelberg (202) 566-0830

Title	Primary Objective	Estimated/Actual Start Date
Carryover		
Feasibility of EPA Achieving Its Goal of Reducing the Gulf of Mexico Hypoxic Zone	To determine to what extent the EPA and states in the Mississippi River watershed are reducing nutrients that contribute to the Gulf of Mexico Hypoxic Zone.	January 2013
Discretionary		
EPA Region 2 Oversight of U.S. Virgin Islands Authorized Environmental Program	To determine how Region 2 oversees the U.S. Virgin Islands authorized environmental program to ensure that they effectively protect human health and the environment.	October 2013
EPA's Oversight of Hydraulic Fracturing	To determine whether the EPA is effectively and efficiently managing the environmental and health risks to drinking water and surface water.	November 2013
Retrospective Study of Drinking Water State Revolving Fund Loans	To determine whether Drinking Water State Revolving Fund loans are issued only after a public water system has demonstrated it has the technical, managerial and financial capacity to operate.	November 2013
Municipal Separate Stormwater Sewers: Consent Decree Progress and Challenges	To determine what results the major municipal stormwater improvement programs had on compliance and environmental quality.	January 2014

Land Cleanup and Waste Management

The Land Cleanup and Waste Management product line is responsible for conducting evaluations to assess EPA programs, activities and initiatives to protect human health and the environment through cleanup and waste management, accident prevention and emergency response.

Point of Contact: Tina Lovingood (202) 566-2906

Title	Primary Objective	Estimated/Actual Start Date
Carryover		
Review of Office of Solid Waste and Emergency Response Cross Program Revitalization Measures	To determine whether the EPA's designation of assessed and cleaned-up sites that have achieved the "ready for anticipated uses" and/or "protective for people" performance measures include effective controls.	April 2012
Hazardous Waste Discharge by Publicly Owned Treatment Works	To evaluate the effectiveness of the EPA's programs in preventing and addressing contamination of surface water from hazardous wastes passing through Publicly Owned Treatment Works.	March 2013
Cross Program Revitalization Measures – Hyperspectral Imaging in Region 4	To determine whether hyperspectral imaging data is a useful tool for assessing contamination and cleanup at Brownfields, Resource Conservation and Recovery Act, Leaking Underground Storage Tank, and Superfund sites.	April 2013
Human Exposure from Lead Smelters	To determine what the EPA has done to address the sites in its Lead Smelter Strategy.	June 2013
Discretionary		
EPA Region 2 Oversight of U.S. Virgin Islands Authorized Environmental Program	To determine how Region 2 oversees the U.S. Virgin Islands authorized environmental program to ensure that they effectively protect human health and the environment.	October 2013
EPA Oversight of the Import of Hazardous Waste	To determine whether EPA oversight of the import of hazardous waste is accomplishing the identified goals.	November 2013
Environmental Risks from Resource Conservation and Recovery Act Hazardous Waste Post-Closure Landfills	To determine the public health, environmental and fiscal risks associated with the expiration of the 30-year post-closure time period.	November 2013
Siting Renewable Energy on Potentially Contaminated Land and Mine Sites: Environmental, Health and Financial Risks	To determine whether the EPA's efforts to promote siting renewable energy on potentially contaminated land and mine sites ensure short- and long-term human health protection.	December 2013
Effectiveness of Third Party Certifications in State-Led Superfund Cleanups	To determine how effectively third party certifiers for state-led hazardous site cleanups are being used by states to address the backlog of contaminated sites.	December 2013

Toxics, Chemical Management and Pollution Prevention

The Toxics, Chemical Management and Pollution Prevention product line is responsible for conducting evaluations to assess the EPA's management of chemical risks and programs to prevent pollution.

Point of Contact: Jeffrey Harris (202) 566-0831

Title	Primary Objective	Estimated/Actual Start Date
Carryover		
EPA's Laboratory Fraud Prevention	To determine the use of procedures by the EPA, states and other federal agencies to manage the communication of and appropriate action on laboratory data determined to be fraudulent.	August 2012
EPA's Greener Product Programs – Conventional Reduced Risk Pesticide Program	To determine whether the Conventional Reduced-Risk Pesticide Initiative is meeting its goal of reducing risks to human health and the environment by encouraging the development, registration and use of pesticide products that are lower risk.	January 2013
Discretionary		
Design for Environment Partnership Program	To determine how effective the EPA's design is for the environment labeling program highlighting safer products for consumer use.	November 2013
Adequacy of EPA's Oversight of State Federal Insecticide, Fungicide and Rodenticide Act Programs	To determine the efficiency of the EPA's oversight of the states' implementation of the Federal Insecticide, Fungicide and Rodenticide Act program.	October 2013
EPA's Use of/Adherence to Quality Management Policies	Determine to what extent the EPA's Office of Pollution Prevention and Toxics' Risk Assessment Division uses and implements policies during chemical risk assessments.	October 2013
National Pesticide Information Center Federal Insecticide, Fungicide and Rodenticide Act Programs Enforcement Referrals	To determine whether Federal Insecticide, Fungicide and Rodenticide Act and pesticide misuse issues that have been reported to the National Pesticide Information Center are being adequately resolved by federal or state authorities.	April 2014

Science, Research and Management Integrity

The Science, Research and Management Integrity product line conducts independent evaluations of EPA's research and development programs and operations managed and directed by the Office of Research and Development. Particular focus is given to those areas that support human health and environmental protection. The product line also develops, coordinates and reports on OIG-identified agency management challenges and internal control weaknesses.

Point of Contact: Patrick Gilbride (303) 312-6969

Title	Primary Objective	Estimated/Actual Start Date
Carryover		
OIG Hotline Complaints – Management of Travel and Trust Funds in Region 6 Water Quality Protection Division	To determine whether the division used trust funds in accordance with applicable federal laws and regulations as well as any agreement with the U.S. Army Corps of Engineers.	February 2013
Discretionary		
Effectiveness of Controls over the Sustainable and Healthy Communities Research Program	To determine the effectiveness of internal controls over the Office of Research and Development's Sustainable and Healthy Communities Research.	March 2014
EPA's Use of Other Federal Agencies, Universities and Foundations for Research	To determine the extent to which the EPA utilizes external sources for agency research and to identify controls that could enhance/support the EPA's use of external sources to meet program objectives.	May 2014
Equipment Utilization Within the Office of Research and Development	To determine whether the Office of Research and Development has adequate controls over research equipment, including utilization, maintenance safeguarding and calibration.	June 2014
Mandated		
Management Challenges and Internal Controls Weaknesses for 2014	To provide the Administrator and Congress those issues which present the greatest challenge to the EPA.	January 2014

Special Program Reviews

The Special Program Reviews product line is responsible for conducting evaluations to assess agency programs and functions to determine whether sufficient controls are in place to reduce the agency's risk of fraud, waste and abuse in its operations.

Point of Contact: Eric Lewis (202) 566-2664

Title	Primary Objective	Estimated/Actual Start Date
Carryover		
EPA Controls Over Time and Material Contracts	To determine whether the EPA processes and procedures require verification that contractor personnel have the qualifications and credentials specified in the contract.	May 2011
Alternative Asbestos Control Method Special Review	To assess the EPA's management controls for the Alternative Asbestos Control Method experiments.	February 2012
STAR (Students to Achieve Results) Grant Hotline	To review a hotline complaint allegation.	July 2012
Discretionary		
Evaluate EPA's Progress Under Environmental Justice Plan 2014	To assess the effectiveness of the environmental justice 2014 plan.	October 2013
Follow Up – EPA Inaction in Identifying Hazardous Waste Pharmaceuticals May Result in Unsafe Disposal	To determine whether the EPA has established a process to review pharmaceuticals for regulation as a hazardous waste and develop an outreach and compliance assistance plan.	October 2013
Follow Up – Weaknesses in EPA's Management of the Radiation Network System Demand Attention	To assess whether the EPA established and enforced expectations for Radiation Network System operations readiness.	October 2013
Review of EPA's Antimicrobial Testing Program	To determine whether the EPA needs to consider upgrades to its antimicrobial testing program that stakeholders have stated is ineffective.	January 2014
Mandated		
Review of EPA's Classification of National Security Information	To assess the EPA's self-inspection program of its National Security Information program.	August 2014

Office of Investigations

The OIG's Office of Investigations (OI) primarily employs criminal investigators (Special Agents), as well as computer specialists and support staff. OI maintains a presence in most EPA regions and at selected EPA laboratories, other facilities and headquarters. The majority of investigative work is reactive in nature.

OI receives hundreds of allegations of criminal activity and serious misconduct in EPA programs and operations that may undermine the integrity of, or confidence in, programs, and create imminent environmental risks. To prioritize its work, OI evaluates allegations to determine which investigations may have the greatest impact on agency resources and on the integrity of an EPA program and operation, and produce the greatest deterrent effect. OI contributes to EPA's strategic goals by ensuring that the agency's resources are not pilfered by criminal activity or criminals.

OI has identified the following major areas on which to focus its investigative activity:

- Financial fraud (contracts and assistance agreements).
- Threats directed against EPA employees, facilities and assets.
- Alleged criminal conduct or serious administrative misconduct by EPA employees.

OI supports the agency and conducts OIG oversight and assistance, as directed by statute and OMB, by providing fraud awareness, detection and prevention training to federal, state, tribal and local officials. OI manages the EPA OIG Hotline Program, which receives hundreds of complaints, referrals and allegations of abuse and misconduct. Additionally, OI is responsible for identifying and investigating attacks against the EPA's computer and network systems to protect resources, infrastructure and intellectual property.

Point of Contact: Patrick Sullivan (202) 566-0308

Investigations begun prior to FY 2014 and new investigations will examine:

- Criminal activities and fraud in programs funded under the American Recovery and Reinvestment Act (Recovery Act).
- Criminal activities in the award, performance and payment of funds under EPA contracts, grants, and other assistance agreements to individuals, companies and organizations.
- Contract laboratory fraud relating to water quality and Superfund data, as well as payments made by the EPA for erroneous environmental testing data and results, that could undermine the bases for EPA decisionmaking, regulatory compliance and enforcement actions.
- Criminal activity or serious misconduct affecting the integrity of EPA programs that could erode the public trust.
- Threats directed against EPA employees, facilities and assets.

- Intrusions into and attacks against the EPA's network, as well as incidents of hijacking EPA computers and/or systems in furtherance of criminal activities, and use of outside computers to commit fraud against EPA.
- Alleged criminal conduct or serious administrative misconduct by EPA employees.
- Disaster relief spending, including participating with other federal OIGs and the EPA OIG Office of Audit on the Hurricane Sandy Fraud Taskforce.
- Small Business Innovative Research grant fraud proactive investigative projects.
- Fraud indicators at a Superfund site in New York City.

OI will continue fraud awareness briefings and training of key EPA officials and other stakeholders to increase their awareness of the indicators of contract and grant fraud and to identify and report funds at risk, as well as recognize and refer cyber threat issues and indicators of vulnerabilities.

OIG Assignments Planned for CSB

The U.S. Chemical Safety and Hazard Investigation Board (CSB) was created by the Clean Air Act Amendments of 1990. The CSB's mission is to investigate accidental chemical releases at facilities, report to the public on the root causes, and recommend measures to prevent future occurrences.

In FY 2004, Congress designated the EPA Inspector General to serve as the Inspector General for the CSB. The OIG has the responsibility to audit, evaluate, inspect and investigate the CSB's programs, and to review proposed laws and regulations to determine their potential impact on CSB programs and operations. During FY 2014, the OIG plans to assess the following for CSB:

- Does CSB provide timely, accurate, complete and useful information for decisionmaking?
- Are CSB programs and operations performing with the greatest efficiency and effectiveness in regard to allocation and application of resources?
- Are the CSB's computer security and privacy programs comprehensive and actively implemented throughout the organization to balance risk and mission requirements?

Title	Primary Objective	Estimated/Actual Start Date
Carryover		
CSB FY 2013 Financial Statements Audit (Contracted)	To monitor contractor to complete audit of FY 2013 financial statements.	October 2012
CSB Contracts	To examine whether the CSB effectively manages its support contracts.	June 2013
CSB FY 2013 Federal Information Security Management Act Audit (Contracted)	To conduct an independent audit of the CSB's compliance with the Federal Information Security Management Act.	August 2013
Mandated		
CSB Improper Payments Elimination and Recovery Improvement Act Compliance Review – FY 2014	To determine whether the CSB is compliant with the Improper Payments Elimination and Recovery Improvement Act.	October 2013
FY 2014 CSB Financial Statements (Contracted)	To monitor contractors to complete audit of FY 2014 financial statements.	June 2014
FY 2014 Management Challenges and Internal Control Weaknesses for CSB	To develop the OIG input to the CSB on FY 2014 Management Challenges.	March 2014
CSB – FY 2014 Federal Information Security Management Act Audit	To conduct an independent audit of the CSB's compliance with the Federal Information Security Management Act.	June 2014

Appendix A—Performance Measures and Targets

The Government Performance and Results Act requires federal agencies to develop goal-based budgets supported by annual performance plans that link the organization's mission and strategic goals to its annual performance goals. The annual performance goals are quantifiable targets supported by measures and indicators representing the expected outputs and outcomes. The agency's annual Performance Accountability Report includes actual results compared to targets to inform OMB, Congress and the public about the value they are receiving for funds invested and how well the OIG is achieving its goals.

This annual plan explains how the OIG will convert its resources into results and benefits of its work through required and priority assignments. Outcome results and benefits from OIG work reflect measurable actions and impacts, but there is typically a time lag between the completion of OIG work and recognition of such results and benefits. Therefore, results and benefits from OIG audits, evaluations, investigations and reviews are recorded in the year they are recognized regardless of when the work was performed. Through current-year outputs and long-term outcomes, OIG targets and seeks to measure and demonstrate the many ways the OIG promotes economy, efficiency and effectiveness; and prevents and detects fraud, waste and abuse. The following are the OIG annual performance goals that this plan is designed to achieve, pending final budget agreements:

Annual performance measures	Supporting indicators	FY 2014 targets (based upon Pres. Budget funding level)
Environmental and business actions taken for improved performance and reduction of risk from or influenced by OIG work.	○ Policy, process, practice, or control changes implemented. ○ Environmental or operational risks reduced or eliminated. ○ Critical congressional or public concerns resolved. ○ Certifications, verification, or analysis for decision or assurance.	307 total
Environmental and business recommendations or risks identified for corrective action by OIG work.	○ Recommendations or best practices identified for implementation. ○ Risks or new management challenges identified for action. ○ Critical congressional/public actions addressed or referred for action. ○ Outreach/technical advisory briefings.	786 total
Potential monetary return on investment in the OIG, as a percentage of the OIG budget.	○ Recommended questioned costs. ○ Recommended cost efficiencies and savings. ○ Fines, penalties, settlements, restitutions.	125% return on investment of budget
Criminal, civil, administrative, and fraud prevention actions taken from OIG work.	○ Criminal convictions / Civil judgments. ○ Indictments/informations. ○ Administrative actions (staff actions and suspension or debarments).	90 total

Appendix B—Risks, Priorities and Issues Identified by OIG During EPA Outreach Interviews With Agency Management

The OIG is highly committed to being a customer-driven organization that provides products and services that address the needs and concerns of agency management. Our planning processes are highly dependent upon, and reflective of, the input received through our outreach to the agency. A summary of current identified areas of concern from the agency is provided below. This information is used by staff as a foundation to lead to the selection of well-supported assignments that answer compelling needs with measurable results.

EPA Cross-Cutting Risks	EPA Outreach Interviews Areas of Concern
Emergency Preparedness/ Homeland Security	• Preparedness for emergencies (natural or manmade disasters) is an unknown risk and needs greater attention. In addition, EPA needs to continue to mitigate the past and future impacts of disasters. • Protection of drinking water from emerging contaminants (Water Sentry program) requires a coordinated effort. • Waste management under possible disaster conditions presents a secondary risk that needs attention. • Data security and protection controls may be vulnerable and should be tested to guard against cyber attack. • Clarification of roles and responsibilities (within the EPA, and between federal agencies and states) needs to be determined and articulated for better collaboration. • The need for a statute on how we deal with imports (with possible health impacts on citizens) is needed to ensure emergency preparedness/homeland security.
Better Collaboration/Coordination With States and Other Federal Agencies	• The 30 federal agencies with an environmental mission need better coordination in planning and implementation. • There is a lack of direct lines of authority (coordination) among and between Assistant Administrators and regions. • Plans, resources, data, authority and measures are not aligned with risks and priorities across the EPA. • Better collaboration internally and with stakeholders is needed to align processes, leverage resources, implement controls, reduce duplication, examine best practices and align resources with priorities. • The EPA needs to coordinate with Department of Homeland Security for streamlined efforts on the new President Directive on Cyber Security for Water Security. • Oil and gas issues on tribal land complicate environmental issues and require better collaboration. • Gulf Coast restoration requires collaboration and coordination with states and other federal agencies.

EPA Cross-Cutting Risks	EPA Outreach Interviews Areas of Concern
Consistent and Reliable Data and Performance Measurement	• There are gaps and inconsistencies in the information that drives the decisionmaking process. • Questions exist as to whether the EPA is collecting the right data, of sufficient quality, and is making that data available. The agency needs to examine the quality of performance measures to ensure activities are properly compiled. • The EPA's information systems are not aligned for efficiency, consistency, accessibility and security. • Control of laboratory data, personally identifiable information and confidential business information outside of EPA, especially related to registration and re-registration of pesticides and other formulas regulated by the Toxic Substances Control Act, all present significant risks. Improvements to data quality from contract laboratories are needed. • Clean Water Act standards are measured differently in each state so information collected is not consistent. • Better quality data is needed from multiple data points to ensure consistent and reliable information.
Improving EPA Organizational Design and Coordination of Resources to Eliminate Duplication	• EPA and its partners need a clear linkage among goals, resources, processes, actions taken and outcomes. • There are no standards or agreements among stakeholders on which to base measures of environmental risks and outcomes (states vs. national). • Program efficiency, progress and results are not measured meaningfully. • EPA does not know what activities cost and what efficiency measures are needed. The agency lacks information needed to assist with determining when investments need to be made in relation to other priorities. • Existing statutes are very prescriptive and allow limited flexibility in managing compliance. Many statutes may not be relevant today and revision may be needed to comply with existing high risk areas. • Differences exist in the ways environmental laws are monitored and enforced between the EPA and states/tribes. Monitoring requirements for grants are underfunded. • EPA must streamline administrative functions to eliminate unnecessary redundancy.

EPA Cross-Cutting Risks	EPA Outreach Interviews Areas of Concern
Monitoring of States, Grants Management, Compliance and Enforcement (How Much Delegation? Federal vs. State Roles?)	• EPA lacks control of fund management and accountability once the funds for assistance agreements to grantees are distributed; half of the agency's budget is allocated to these agreements. • The highest risk in the grants management process is at the point that funds are spent by grantees and are sometimes commingled with other sources of grant funds. • Grantees have limited capacity or incentive to account for funds or performance. • The EPA lacks resources to adequately monitor grants and lacks uniform reporting and accountability conditions. • The EPA should execute and manage grants for measurable success vis-à-vis their intended goals. • The EPA needs to determine how to get the best balance for return on investment between mandatory and voluntary actions.
Human Capital Management – Skill Gaps/Alignment With Functions	• The EPA should analyze its workforce to identify and fill skill gaps and to implement its Human Capital Strategy. • The EPA needs to determine programs and areas that can be done locally versus nationally to decrease overhead. • The EPA must determine whether employees in its workforce are aligned in the right places.
Better Use of Technology, Information and Research	• The EPA should manage its resources and the performance of contractors to optimize their value added. • The EPA needs operational controls to protect and account for costs, assets, information and performance. • The EPA should more strongly implement FMFIA and the OMB Circular A-123 process. • The Working Capital Fund lacks the transparency or accountability necessary to prove its efficiency. • Agency management should better understand and be accountable for taking agreed-to actions on OIG recommendations.
EPA's Regulatory Process (Better and Faster Analysis of Costs, Science and Benefits)	• The EPA's extremely complex regulatory process should be streamlined without compromising its required integrity. • Competing interests of stakeholders and the regulated community may lead to overlaps, gaps and conflicts. • Many policies are out of date or are based on outdated science and technology. • EPA should evaluate how to use voluntary incentives for compliance.

EPA Cross-Cutting Risks	EPA Outreach Interviews Areas of Concern
Cross-Media Risk Assessment, Planning and Priority Setting for Better Application of Resources	• The EPA should use a consistent approach to evaluate actual and relative environmental and operational risk and program effectiveness, assign resource priorities, make regulatory decisions, take enforcement actions, and inform its stakeholders. • EPA should ensure the integrity of laboratory data, results and scientific research; knowledge and innovative technology should be transferred in a timely manner in the regulatory and policy process. • Agency programs need a consistent approach for determining relative risk and demonstrating outcome results.
Water Infrastructure, Financing and Water Availability	• The EPA needs to address failing infrastructure for drinking and storm water systems. Approximately $20 billion will be needed to stabilize infrastructure across states. • It is unclear who will pay for needed infrastructure investment. • Hydro fracking in New York needs a before-and-after study. • EPA should examine how natural gas should be regulated under the Clean Water Act.
Land and Superfund	• It appears that Superfund sites are taking an extraordinary long time to address. The agency needs to address this issue and determine whether management issues are preventing sites from doing cleanups. • The EPA needs to examine chemical safety and ensure that states are monitoring this problem to ensure safety of communities.
Climate Change and Air	• EPA should determine how to use creative financing and leverage funding through public/private partnerships. • EPA should utilize a better method for understanding air toxics and their monitoring. • EPA needs a clear and unified strategy, including participation of other federal agencies and other national governments. • Climate change in the northeast needs to be analyzed and determine why rebuilding always focuses on same places.

www.ingramcontent.com/pod-product-compliance
Lightning Source LLC
Chambersburg PA
CBHW081410170526
45166CB00010B/3291